T0200235

John Lewis-Stempel is a writer and farmer. His previous books include *The Wild Life: A Year of Living on Wild Food*, *Meadowland: The Private Life of an English Field*, which won the 2015 Wainwright Prize, *The Running Hare: The Secret Life of Farmland*, which was shortlisted for the 2017 Wainwright Prize, *Where Poppies Blow*, which won the prize, *The Wood: The Life and Times of Cockshutt Wood*, and most recently *Still Water: The Deep Life of the Pond*. John writes for *Country Life* and won the BMSE Columnist of the Year Award in 2016. He lives on the borders of England and Wales with his wife and two children.

The
Private Life
of the
Hare

John Lewis-Stempel

doubleday

TRANSWORLD PUBLISHERS
61–63 Uxbridge Road, London W5 5SA
www.penguin.co.uk

Transworld is part of the Penguin Random House group of companies
whose addresses can be found at global.penguinrandomhouse.com

First published in Great Britain in 2019 by Doubleday
an imprint of Transworld Publishers

A CIP catalogue record for this book
is available from the British Library

ISBN 9780857524553

Typeset in 11/14.5pt Goudy Oldstyle Std
by Integra Software Services Pvt. Ltd, Pondicherry

Printed and bound in Great Britain by Clays Ltd, Elcograf S.p.A.

Penguin Random House is committed to a sustainable
future for our business, our readers and our planet. This book
is made from Forest Stewardship Council® certified paper.

7 9 10 8

For my grandfather, Joe Amos, farmer
and lover of the hare

The Hare

In the black furrow of a field
I saw an old witch-hare this night;
And she cocked a lissome ear,
And she eyed the moon so bright,
And she nibbled o' the green;
And I whispered 'Whsst! witch-hare',
Away like a ghostie o'er the field
She fled, and left the moonlight there.

Walter de la Mare (1873–1956)

CONTENTS

PROLOGUE

My First Hare

LONG AGO THERE was a TV show for children called *Play School*, in which the audience at home were invited to look through either the square, round or arched window at something interesting, or at least deemed noteworthy by adults. The programme hit a particular button with the young me because the house in which I lived, despite being ancient and stone, had an iron-and-glass porthole in the upstairs bathroom. This incongruous architectural feature had been installed in the 1950s by a retired Royal Navy commander who clearly wished the house to be shipshape.

There was no ocean outside the porthole, only a sea of crops or of earth, depending on the season. The house was in east Herefordshire, as landlocked as it's possible to get in Britain.

One spring morning, when the dawn sun reddened the world, I stood on a chair and looked out of the bathroom porthole at the field. I was seven or so, and was half-heartedly washing my face with a flannel.

Out of the lapping waves of ploughed clay, a hare suddenly sat up. Since the house was on a steep slope, the hare was almost at eye level. I had seen hares in the distance; and I had seen hares dead, because one of my grandfather's favourite dishes was jugged hare, and in his cold larder shot hares were often to be found hung up, their eyes bulging, their teeth in the rictus snarl of death. Well, we are all diminished by the extinguishing of life.

But I had never seen a live hare at five paces. The hare performed a toilette of licking its paws, as if it too was getting ready for school. I remember thinking that the hare seemed to take pleasure in washing: that it knew a simple state of joy. For maybe five minutes the big hare and the small boy washed on opposite sides of the glass.

Something alarmed the hare, and it bolted. No longer was the hare an enchanting friend; it became absolute motion, a process in physics, a kinetic event. Every cell of the hare was energy as it travelled across the earth.

Hares have no home. Rabbits when threatened make for their burrows, blackbirds dive into thickets. Hares run fast for their lives.

I have seen many hares since, but the hare in the wheatfield at Withington was the one which showed me that hares, unique among the animals, can alter state.

INTRODUCTION

H<small>E RAN LIKE</small> a hare, we say. Timid as a hare. To 'run with the hare and hunt with the hounds' is to keep favour with both sides. 'Hare-brained' is a mind that is daffy, skeetering about in all directions; the phrase dates from 1542. Shakespeare depicted the impetuous Henry Percy as 'A hare-brain'd Hotspur, govern'd by spleen'.

Our fascination with hares is old, and boundless. One proof is the hare's impact on our lexicon, another is that on our culture. In the Hare and Tortoise fable, Aesop paid the hare's superlative speed a back-handed compliment, while the Br'er Rabbit stories of the American South credit the hare's cunning. (The Americans, in a typical piece of 'I say tomato, you say tomayto' confusion, call the hare 'Br'er rabbit'.) The contemporary Easter Hare who dispenses gifts to children, usually chocolate eggs, is an echo of a prehistoric fertility ritual. Lewis Carroll created the most famous modern image, the Mad March Hare who attends the tea party in *Alice's Adventures in Wonderland* (1885) – where the hare is both absurd and minatory: 'The March hare will be much the most interesting,' reasons Alice, 'and perhaps, as this is May, it won't be raving mad – at least not so mad as it was in March.'

Hares do *seem* mad in the month of March, when they 'box' in fields. ('Mad as a March hare' is yet one more entry in the phrasebook of hare-birthed expressions.) For centuries it was assumed that the springtime fighting of hares was between males; only in the last twenty years has it been proved that the pugilist who takes on all comers is a female fending off unwanted males. But hares, almost more than any other European mammal, have been misunderstood. Pliny in the *Natural History* wrote that alpine hares turn white because they eat snow and that eating hare flesh 'would causeth them that feed upon it to look fair and gracious for a week afterwards'. As late as 1607, Edward Topsell informed readers of his *History of Four-footed Beasts* that the hare is 'one year male, and another female'.

There is good reason for the historical falsehoods about the hare. Hares, which tend to be most active at night, are secretive, remote, thus unknowable. In the absence of observed facts, speculation and fantasy flourished. The lean beast became girded with weird beliefs. The hare was a shape-shifter, a familiar of witches. Even today, when formerly rural animals such as the fox have taken up residence *in urbe*, the hare, an irredentist country-dweller, is a rare sight for most people. In an age when animal after animal has lost its magic, the hare runs along the cusp of reality. To see a running hare is to peep into other worlds.

But real hares? What are they like, truly?

I

The Private Life
of the Hare

A HARE IS not a rabbit. There are easily observable physical differences between the two British members of the taxonomic order *Lagomorpha*. The hare has the ears of a donkey, the golden eyes of a lion, the chiselled face of a horse, and the legs of a lurcher. The Ming Chinese considered the hare so other-worldly in its assemblage of bits from other fauna they decided its ancestor lived in the moon. The rabbit is altogether more ordinary, with shorter ears, nondescript brown eyes, the sort of sweet face that enables it to become the 'bunny' toy of childhood.

You remember your first close encounter with a hare, in the way you remember your first swim in the sea, your first trip to the cinema.

The hare's distinctive legginess can deliver speeds of 48mph, making it the fastest land mammal in Britain. In a single bound, a hare can cover ten feet, and if pursued can jump a Grand National fence. The scientific name of the hare is *lepus*, from the Latin *levipes*, 'light foot', because of speed and agility such as this. The expression 'to kiss the hare's foot' means to be late, alluding to the hare's great speed and the fact that, if you hesitate, it will be gone and all that will be left is a footprint.

Unlike a rabbit, the hare's top lip is split to the nose ('hare-lip'), and in aged hares the teeth will, walrus-like, protrude through the gap.

The difference between hares and rabbits is more than skin deep. A hare's heart is very large, weighing between 1 per cent and 1.8 per cent of the body weight; a rabbit's heart weighs about 0.3 per cent of body weight. The hare's super-sized heart is the secret engine which powers the animal's elegant limbs. The hare's broad, brown nose contains extensive turbinate bones, thin and delicate as glass, which maximize oxygen – hare booster fuel – intake.

Although lagomorphs are a venerable order, one almost unchanged since the Pleistocene period, it is believed that the brown hare, the commonest hare in

Britain, spread from the Russian steppes into northern Europe after the last Ice Age. Probably the Iron Age people introduced the brown hare to Britain, and this species over time pushed the native mountain hare into Scotland.

All hares rely on the cryptic colouration of their fur for protection since, unlike rabbits, they are surface dwellers, not burrowers. By day, hares sleep and shelter close to the ground in grass or earth depressions known as 'lairs' or 'forms'. Forms are carefully selected, to give shelter but also a view and a 'nose', so tend to be found on dry, elevated aspects with the hare's head facing into the prevailing wind. The hare's eyes are set far back on the head, which enables them to monitor a circumference around them of almost 360 degrees. Though popularly supposed to sleep with their eyes wide open, hares do close their eyes when they fall into semi-sleep. Deep sleep lasts for a mere minute or three. The eternally alert, quick hare became a Christian symbol of vigilance and the need to flee from sin and temptation.

Brown hares, disguised to resemble earth, are wholly confused by snow, as Gilbert White noted in an entry for January 1776 in *The Natural History of Selborne*: 'The hares also lay sullenly in their seats, and would not move till compelled by hunger; being conscious, poor animals, that the drifts and heaps treacherously betray their footsteps, and prove fatal to numbers of them.'

As far as hares are concerned, snow is an enabling highlighter for predators. Hares, in snow, will lie buried save for a tiny blowhole near the nose for up to six days. If a wood is nearby, they might retreat to its sheltering edge.

Those long ears? They can swivel 270 degrees plus. And scope like the ear-trumpets of Edwardian aunts.

After camouflage and sight, the hare's next tactic for avoiding enemies is flight. Only when detection by a predator is certain, when the nose of a dog or the foot of a human is almost upon it, will a hare run for its life. Once, when I was young, I walked through a hay meadow near Tenbury Wells and a hare burst from under my boot, like a mammalian explosive. Until that moment the grass bulge had been a simple tussock; the hare shot from its stalky den with the speed and purpose of a torpedo.

The hunted hare will instinctively head uphill, using to its advantage its muscular hind legs which are

several inches longer than the fore limbs. ('Muscular' barely covers the superpower physicality of a hare's back limbs. I've held a wild hare: I might as well have tried to contain the whirlwind.) At top speed, the ears will lie flat for streamlining. The running hare, who can see the pursuant enemy because of those generously proportioned eyes, is full of tricks: it will 'maze' or swerve, and for those predators pursuing it by scent, it will lay out a repertoire of doubling back and springing in order to confuse.

Shakespeare, no mean naturalist, noted this wily tactic of the hare in his long poem *Venus and Adonis*:

Sometimes he runs among a flock of sheep,
To make the cunning hounds mistake their smell,
And sometime where earth-delving conies keep,
To stop the loud pursuers in their yell,
And sometime sorteth with a herd of deer:
Danger deviseth shifts; wit waits on fear . . .

For there his smell with others being mingled,
The hot scent-snuffing hounds are driven to doubt . . .

A hare running is poetry in motion; but a hare hopping around the cabbage patch in search of food is prose, gauche and ungainly. The hare's reliance on speed means it is essentially a creature of open spaces – the field, moor and heath.

Perversely, the terrestrial hare, as if to resist easy categorization, can swim well. The Victorian naturalist the Revd E. A. Woodruffe-Peacock watched a hare swimming the River Trent, at a point where the river is over two hundred yards wide, in order to feed off carrots.

Usually, the hare emerges from the sanctuary of its form at twilight in order to eat; the female, a 'doe' or a 'jill', is 60–70cm long and is slightly larger than the male, the 'jack' or 'buck'. Hares are strictly vegetarian, and need a rich diet of plants and herbs. With 95 per cent of hay meadows gone in favour of species-poor silage, brown hares currently fare best in the arable areas of eastern England.

In the Old Testament book of Deuteronomy it was ruled: 'Nevertheless these ye shall not eat of them that chew the cud, or of them that divide the cloven hoof; as the camel, and the hare, and the coney: for they chew the cud, but divide not the hoof.' Hares do not quite chew the cud like ruminants, who regurgitate their food, chew it again and then swallow it again. In lagomorphs, an offshoot of the gut (the caecum) acts as a fermentation chamber to break down the cellulose which grasses contain. The caecum then produces soft, lozenge-shaped faeces which the lagomorphs re-ingest to improve the bacterial efficiency of digestion and extract additional nutrition. After a second passage of material through the gut the familiar currant-like pellets are produced. This behaviour is known as caecotrophy or refection. In other words, hares eat their own excrement. Despite this neat nutritional trick, hares eat for hours a day; by my observation on a day of sunlight and moonlight in April, for at least twelve hours. Over the course of a night a hare might cover ten miles in its gastronomic seekings; brown hares are recorded as eating more than sixty-three types of plant. A hare in winter needs quantity as well as quality: 2,800 kJ per day.

Usually as solitary and lonely as a hermit, the hare will sometimes gather in groups in winter to feed. This is particularly true of the mountain hare, which will form a gang of twenty or more on the leeward side of a hill, scraping away shallow snow in order to feed.

The hare's life of solitude is also broken for mating. Hares breed most of the year, and in mild years they may breed the year round. The kangaroo-like boxing ritual of hormonal hares is most clearly seen in March and early spring when the crops are low and the light is relatively good. As 'rituals' go, the boxing of hares can be decidedly violent, with fur flying; sometimes a kicking and swiping doe will rip a male's ears so they spray the morning with rain-blood.

However, hares are at maximum fertility in the spring, with this prime breeding season triggered by the lengthening of daylight. The majority of does first become reproductively active in their second calendar year, when they have reached a weight of around 1.5kg. Three litters a year is normal, four litters is not exceptional. Usually, litters consist of two 'leverets', as hare young are known. Gestation is between forty and fifty days.

Some early writers on natural history believed that the hare had a separate womb for each leveret. Claudius Aelianus in the third century AD wrote that the hare 'carried some of its young half formed in the womb . . . others it has already borne'. Aelianus was groping towards one of the oddest biological truths about the hare: it is capable of superfoetation, where the new eggs of an already pregnant doe can be fertilized, giving a potential minimum duration between successive litters of thirty-eight days.

Newly born leverets, which are the cute poster-pic animals of British wildlife, are about 100g in weight, fully furred, have open eyes, and receive little parental care other than suckling visits by their mother. (Rabbit young, by contrast, are born bald, with their eyes closed.) A day or two after birth the doe, as a precaution against predators, will move the leverets from the birth form, and place each leveret in its own form. The naturalist Brian Vesey-Fitzgerald was convinced that the doe carried the leverets to their new form in her mouth, 'after the manner of a cat carrying kittens'. Other observers believe the leverets walk to their new forms, since they are perfectly capable of doing so within a day of birth.

Until they are weaned their mother visits each leveret in turn at or just after sunset. The doe will 'mew' to the leverets as a signal that she has taken up her feeding station. They can also scent her. Since hare milk is very rich, suckling takes only a minute or two. While the leverets suckle the doe is wired with alertness; she will hear, with those super-size black-tipped ears, a footfall two fields away. A momentary suckling once a day may seem the absolute minimum that a doe could do for her young, but the less she visits the less chance there is for her to lead a predator to them.

When frightened, leverets press tight to the ground and become motionless. Impersonating a clod of earth or a stone is not always sufficient or successful. Hares are in danger from the first day of their existence. Predation by the red fox accounts for more than 70 per cent of annual losses of the brown hare, while the golden eagle is a principal killer of the mountain hare.

For all who prey on the allegedly timid hare, there is one final surprise. They will fight back. We had a hare in our wheatfield who jumped up at a swooping

buzzard to give it a blow on the beak. On another occasion, a doe hare reared up from her maternity form to hiss at a cow who was getting too close for safety. And the hare standing in the stubble to eyeball the approaching fox, catching Reynard out in the game of hide-and-seek, is a commonplace.

Hares at Play

The birds are gone to bed, the cows are still,
And sheep lie panting on each old mole-hill;
And underneath the willow's grey-green bough,
Like toil a-resting, lies the fallow plough.
The timid hares throw daylight fears away
On the lane's road to dust and dance and play,
Then dabble in the grain by naught deterred
To lick the dew-fall from the barley's beard;
Then out they sturt again and round the hill
Like happy thoughts dance, squat, and loiter still,
Till milking maidens in the early morn
Jingle their yokes and sturt them in the corn;
Through well-known beaten paths each nimbling hare
Sturts quick as fear, and seeks its hidden lair.

John Clare (1793–1864)

The Hares of Britain and Ireland

Hares exist on every continent, with the exception of Antarctica. Along with rabbits and pikas, hares comprise the scientific order *Lagomorpha*, which differs from rodents in having four incisors in the upper jaw (not two, as in *Rodentia*), having a scrotum in front of the penis, and being wholly herbivorous.

We have three species in Britain and Ireland, the European brown hare, the mountain hare, and the Irish hare.

The Brown Hare

Scientific Name: *Lepus europaeus* (formerly *Lepus capensis*)

Size: 60–70cm, including 7cm tail

Description: Very long ears tipped with black; long, powerful hind legs. Brown furry body, with a reddish and sometimes a golden tinge. Pale under chest. Moults twice per year.

Weight: Average 3–4kg, with specimens of 5kg recorded in the arable lands of East Anglia

Diet: Tender grass shoots, including cereal crops, are their main foods.

Lifespan: Three to four years on average, although one brown hare tagged in Poland lived in the wild for eight years.

Habitat: Widespread on low ground, especially grassland and arable fields, with the exceptions of S.W. England and W. Scotland. Commonest in East Anglia. Although introduced across Northern Ireland by landowners in the nineteenth century, now localized in N.W. Ulster. The Joint Nature Conservation Committee estimates the UK population to be c.817,500.

The Mountain Hare (also known as the blue hare, the white hare, the maukin, the *Maigheach-gheal*)

Scientific name: Lepus timidus

Size: 50–65cm

Description: Long black-tipped ears. Blue-grey fur in summer; in winter, the mountain hare's coat or 'pelage' becomes completely white. The white coat offers insulation, as well as camouflage. Moults three times a year.

Weight: 3–3.6kg

Diet: Grasses, shrubs, heather. In severe weather it will gnaw bark above the snow levels.

Lifespan: Three years on average.

Habitat: Lives in higher latitudes than the brown hare. Native to Scotland, where it has been known to breed as high as 1,200m, with small introduced populations in the Peak District and the Isle of Man. This is our original British hare, and was around in mainland Britain in the Palaeolithic age. Mountain hare bones between 114,000 and 131,000 years old have been found in the Joint Mitnor Cave in Devon and in the Thames Valley. As with brown hares, males are slightly smaller, by about 10 per cent, than females. While brown hares can breed the year round, mountain hare leverets are almost always born between March and August.

The Irish Hare (also known as *Giorria eireannach*)

Scientific name: Lepus timidus hibernicus

Size: 60cm

Description: Distinguished from brown hares by a stockier build, shorter ears; the upper surface of the tail may veer towards white or grey (the top of the brown hare's tail is always dark). Fur colouration is generally reddish brown in summer, grey-brown in winter, though some Irish hares turn partially white in winter. Rathlin Island, off the north coast of Antrim, is home to the Rathlin 'golden hare', which displays a distinct genetic variation producing individuals with blue eyes.

Weight: 3–3.4kg

Diet: They feed mainly on a variety of grasses but sedges, heather, wild thyme, bilberry and even the shoots of young trees may play an important part in their diet, depending on the habitat.

Lifespan: Three to four years

Habitat: The Irish hare *Lepus timidus hibernicus* is the only species of lagomorph native to the island of Ireland. Carbon dating of cave fossils has shown that hares were present in Ireland as far back as 30,000 BC. Generally regarded as a sub-species of the mountain hare which occurs in the rest of the UK, the Irish hare has been revealed by DNA research to be genetically distinct.

Widely distributed, especially in areas with semi-natural grassland, heath or bog, although the population density may be as low as one per square kilometre.

On Yonder Hill There Sits a Hare

Traditional Irish hunting song

On yonder hill there sits a hare.
Full of worry, grief and care,
And o'er her lodgings it was bare,
Singing ho, brave boys, hi-ho.
And o'er her lodgings it was bare,
Singing ho, brave boys, hi-ho.

Now there came a huntsman riding by,
And on this poor hare he cast his eye,
And o'er the bogs halooed his dogs,
Singing ho, brave boys, hi-ho.
And o'er the bogs halooed his dogs,
Singing ho, brave boys, hi-ho.

And now she's gone from hill to hill.
All for the best dog to try his skill
And kill the poor hare that never done ill,
Singing ho, brave boys, hi-ho.
And kill the poor hare that never done ill,
Singing ho, brave boys, hi-ho.

And now she's turned and turned again,
Merrily as she trips the plain,
And may she live to run again,
Singing ho, brave boys, hi-ho.
And may she live to run again,
Singing ho, brave boys, hi-ho.

(*Roud 5173*)

The Hare's Lament

Irish Traditional

On the first of November on a bright autumn's day
To the hills of Dromela I chanced for to stray
I was feeding on green grass that grows on yon ground
When my heart was set a beating by the cry of the hounds.

Musha right tallyho, hark ye over high ho
Hark ye over cries the huntsman hark ye over high ho.

They hunted me up and they hunted me down
The bold huntsmen of Stratham on my trail sent the hounds.
Over highlands and lowlands moorlands also
Over hedges and ditches like the wind I did go.

There was Ringwood and Rouser they gave me a close brush
But they soon found me hiding twas in the rush bush
For better or worse I know I must die
But I'll do my endeavour these hounds to defy.

And now I must die and I know not the crime
To the value of sixpence I ne'er harmed mankind.
I ne'er was brought up for to rob or to steal
Unless for the croppings some tops of green kale.

Then up steps the huntsman to end all my strife
Saying let the hare go give her play for her life.
Wouldn't it be far better you killed Raymond the fox
Who killed all your chickens, fine hen and game cock?

(Roud 3574)

II

Hunting the Hare

*'Of the hare ... That beest kyng shall be calde of all ven-
ery ... He is the mervellest beest that is in ony londe.'*

The Boke of St Albans (1486)

HARES HAVE BEEN hunted by man from the beginning of painted time. There is an engraved hare
on the wall of the Grotte de Gabillou in the Dordogne
region of France. In all likelihood, the Palaeolithic
artist intended his representation as a magical aid in
the hunt for the hare and the sating of the stomach.

The hare as sporting object was not far behind.
Xenophon, writing in the fourth century BC, penned
a whole treatise, *Cynegeticus* (*On Hunting*), on the
pursuit of hares by dogs; these were hounds, or scent
dogs. In the same Greek times, the constellation

Lepus was said to be the hare running away from the hunting dogs of Orion. Hares were associated with Artemis, goddess of wild places and the hunt, and new-born hares were not to be killed but left to her protection.

The Egyptians, there is evidence, kept hares in special preserves for the purposes of sport-cum-hunting for the pot.

Hare coursing, where 'long dogs' chase down hares in a sight hunt with men on horseback, was an important sport in Roman Gaul. The Christian theologist Father Tertullian even considered hare hunting a sufficient, all-explaining metaphor for the harrying of the Church: 'The hunt is focused on us,' he said, 'as if we were hares.' Arrian, c.AD 180, in his updated *Cynegeticus*, justified the sporting fairness of coursing with sighthounds, in terms much wielded since: 'True huntsmen do not take out their hounds to catch the creature, but for a trial of speed and a race, and they are satisfied if the hare manages to find something that will rescue her.'

Since the Romans introduced the hare to England, it is certain they introduced the sport of coursing. When in Britain (or Gaul, or Asia), the Romans did as they did in Rome. From at least the thirteenth century in England, hares were conserved

and protected by law, especially during the breeding season, in order to be hunted. In Norman
times the hare's speed and trickery displayed in
the hunt earned it the honour of being one of the
four 'Beasts of Venery', together with the hart, the
boar and the wolf. The Beasts of Venery were considered by the nobility to be worthy of the chase,
and honoured accordingly. Conversely, a Norman
coward was personified as an armed man who ran
from the hare.

The English hunter's reverence for the hare con-
tinued into Georgian times (until the lean beast was
replaced in the hunter's pantheon by the fox and
the pheasant). A special breed of fleet hare-chasing
greyhounds was maintained by King John (1166–
1216), at the considerable cost of 109s. (shillings)
every eighteen months. In his treatise on hunting,
The Master of Game (c.1410), Edward, Duke of York,
declared: 'It is to be known that the hare is the king of
all venery; for blowing and the fair terms of hunting
cometh of the seeking and finding of the hare ... It is
a fair thing to slay her with the strength of hounds,
for she runneth strong and cunningly. A hare shall
last well four miles more or less if she be an old male
hare.' (A hare is quite often a 'she' in vernacular;
in East Anglia the hare is still referred to by older
country folk as Owd Sally.) In order to preserve the
hare as a sporting object for the upper classes alone,
laws were enacted. The hare was to be treated with
respect; at Somerton in 1256 twelve jurors sat down
to determine the cause of death of a hare as if it had
been a man. Richard II, who ruled England 1377–99,
passed a statute prohibiting any layman from keep-
ing greyhounds, or catching hares in nets or snares,
unless he could prove that he possessed lands or ten-
ements of the annual value of 40s.

The lawyers of James I's reign framed no fewer than three Acts to repress hare poaching by *plebs sordida*. Of these, the first enacted that, from August 1604, no one might keep a greyhound for coursing hares unless he was either a man of good family or enjoyed an income, in his own or his wife's right, of £10 a year. A measure of popularity was infused into this arrangement with the proviso that the persons who were to benefit from the fines reaped by the enforcement of the law were the poor and needy of the parish. Charles II sanctioned a much more stringent Act, by which the property qualification was raised to £100, due half yearly from landed property, or at least a lease of £150 for ninety-nine years. Only gentlemen of such visible substance, their heirs and keepers were thenceforward to have or to keep guns, bows, greyhounds, hare pipes or other sporting implements.

Laws against the hunting of hares in snow, when they were notoriously easy to track, were particularly draconian. Henry VIII allowed no one, whatever his station in life, to kill a hare in snow. Offenders were fined 6s. 8d. for every hare so killed. James I ordered offenders to prison for six months or to pay 20s. per hare (about £200 in today's money).

The Elizabethans set the rules for the sport of enclosed or park coursing with lurchers and greyhounds. No more than two hounds were allowed for the pursuit, while a head start termed 'Law' was to be given to the hare for a fair run. The dogs were awarded points for 'Speed', 'Go-bye', 'Turn', 'Wrench', 'Kill' and 'Trip'. According to the 1884 by-laws of the National Coursing Club:

> The Judge shall decide all courses upon the one uniform principle that the greyhound which does most towards killing the hare during the continuance of the course is to be declared the winner.
>
> The principle is to be carried out by estimating the value of the work done by each greyhound, as seen by the Judge, upon a balance of points according to the scale hereafter laid down, from which also are to be deducted certain specified allowances and penalties.
>
> The points of the course are:
>
> a. Speed. – Shall be estimated as one, two, or three points, according to the degree of superiority shown.
>
> b. The Go-bye. – Two points, or if gained on the outer circle, three points.
>
> c. The Turn. – One point.

d. The Wrench. – Half a point.

e. The Kill. – Two points, or, in a descending scale, in proportion to the degree of merit displayed in that kill, which may be of no value.

f. The Trip. – One point.

Coursing was introduced into Ireland in 1803 by the British Army.

There were other ways to catch a hare with dogs. Harriers and beagles pursued the hare by means of a good nose and stamina. Hunting hares, with every other variety of hunting wild mammals with dogs, was made illegal in Britain with the 2004 Hunting Act.

Throughout the medieval and Tudor years the smaller mountain hare was the favoured quarry in falconry; the bigger brown hare was a clawful for even a full-grown goshawk.

Although the shotgun was principally developed for killing flying birds, it was inevitable that the hare would be considered 'fair game'. On the shooting estates of East Anglia, where the hare reached such numbers it could reasonably be considered an agricultural pest, huge 'bags' of hares were commonplace. The Stowlangtoft estate game book shows that over the period 1853–1911 41,593 hares were killed on the estate. The British record bag was at Holkham in Norfolk on 19 December 1877 when 1,215 hares were shot.

Shooting hares was always an upper-class hobby. Although poachers might pot a hare with a .410 shotgun, the gentlemen of the night generally preferred long dogs, nets and snares. Hares are creatures of habit, entering and exiting fields by the same 'smeuse' (hole, known in the north of England as 'smout') in the hedge day after day. Richard Jefferies described the standard method of the Victorian poacher in *The Gamekeeper at Home* (1878):

> Though apparently roaming aimlessly, hares have their regular highways or 'runs'; and it is the poacher's business to discover which of these narrow paths are most beaten by continuous use. He then sets his wire, as early in the evening as compatible with safety to himself [from the gamekeeper and magistrate], for hares are abroad with the twilight.
>
> Long practice and delicate skill are essential to successful snaring. First the loop into which the hare is to run his head must be of the exact size. If it be too small he will simply thrust it aside; if it is too large his body will slip through and his hind leg will be captured: being crooked, it draws the noose

probably. Then if caught by the hind leg, the wretched creature, mad with terror, will shriek his loudest;
and a hare shrieks precisely like a human being in
distress ... A hare carries his head much higher than
might be thought; and he is very strong, so that the
plug which holds the wire must be driven in firmly to
withstand his first convulsive struggle.

Jefferies considered that if a poacher snared three
hares a night (wholesale price 2s. 6d. each) 'he is well
repaid'.

Poachers with a gift of imitation were able to
'call' hares into nets and traps by imitating the cry of
a leveret. According to Percy the Poach, the semi-
professional 'moocher of the hedgerow' in my childhood Herefordshire village, hares were susceptible to
being caught by distraction. All that was required was
for an accomplice to walk up behind the hare whist-
ling or throwing a cap, while the perpetrator proper
approached the hare head on and gave it a thwack
from a cudgel. Sometimes hares seem to abandon
all caution, and will lollop, with their distinctive
rocking-chair gait, to look at an intruder.

There has always been sympathy for the hunted
hare, unlike the fox, who could never be portrayed as
an innocent. In the early Tudor period, Thomas More's

Utopians demonstrated a startlingly modern pity for the hare: 'Thou shouldst rather be moved with pity to see a silly innocent hare murdered of a dog, the weak of the stronger, the fearful of the fierce, the innocent of the cruel and unmerciful. Therefore, all this exercise of hunting is a thing unworthy to be used of free men.' Sixteenth-century Michel de Montaigne sighed: 'I cannot endure a dew-bedabbled hare to groan when set upon by hounds.' Montaigne, in turn, influenced Margaret Cavendish, Duchess of Newcastle, who wrote a poem in 1653 entitled 'The Hunting of the Hare':

> *Betwixt two Ridges of Plowd-land sat Wat,*
> *Whose Body press'd to th'Earth, lay close, and squat,*
> *His Nose upon his two Fore-feet did lye,*
> *With his grey Eyes he glared Obliquely;*
> *His Head he always set against the Wind,*
> *His Tail when turn'd, his Hair blew up behind,*
> *And made him to get Cold; but he being Wise,*
> *Doth keep his Coat still down, so warm he lies:*
> *Thus rests he all the Day, till th'Sun doth Set,*
> *Then up he riseth his Relief to get,*
> *And walks about, untill the Sun doth Rise,*
> *Then coming back in's former Posture lies.*
>
> *At last poor Wat was found, as he there lay,*
> *By Huntsmen, which came with their Dogs that way*

Whom seeing, he got up, and fast did run,
Hoping some ways the Cruel Dogs to shun;
But they by Nature had so quick a Scent,
That by their Nose they Trac'd what way he went,
And with their deep wide Mouths set forth a Cry,
Which answer'd was by Echo in the Sky;
Then Wat was struck with Terrour and with Fear,
Seeing each Shadow thought the Dogs were there,
And running out some Distance from their Cry,
To hide himself, his Thoughts he did imploy;
Under a Clod of Earth in Sand-pit wide
Poor Wat sat close, hoping himself to hide,
There long he had not been, but strait in's Ears
The winding Horns and crying Dogs he hears;
Then starting up with fear, he Leap'd, and such
Swift speed he made, the Ground he scarce did touch . . .

Cavendish goes on to describe the horror of the hunt from Wat the hare's point of view. The poem ends with an adjuration to consider the hare a gift from God, and a damning of man as prideful:

Yet Man doth think himself so Gentle and Mild,
When of all Creatures he's most Cruel, Wild,
Nay, so Proud, that he only thinks to Live,
That God a God-like Nature him did give,

41

And that all Creatures for his Sake alone
Were made, for him to Tyrannize upon.

England's first Poet Laureate, John Dryden, was moved, after watching a hare hunt, to write 'Some Fearful Hare':

So I have seen some fearful hare maintain
A course, till tir'd before the dog she lay;
Who, stretched behind her, pants upon the plain,
Past pow'r to kill, as she to get away:

With his loll'd tongue he faintly licks his prey;
His warm breath blows he flix up as she lies;
She, trembling, creeps upon the ground away,
And looks back to him with beseeching eyes.

Robert Burns, in a poem whose title hardly requires parsing, 'On Seeing a Wounded Hare Limp by Me, Which a Fellow Had Just Shot At', excoriated the shooters of hares:

Inhuman man! curse on thy barbarous act,
And blasted be thy murder-aiming eye;
May never pity sooth thee with a sigh,
Nor ever pleasure glad thy cruel heart!

Modern greyhound racing is coursing without the blood, and the pitiful screaming of hares dying. In greyhound racing, the dogs run after an electric 'hare', a rabbit-shaped robot, sometimes scented with meat, and usually with a red light on its rear so the greyhounds can see the lure in the mist.

III

The Hare as Food

'FIRST CATCH YOUR hare,' began Hannah Glasse's famous recipe for jugged hare in *The Art of Cookery Made Plain and Easy* (1747), which is no simple matter when hares run at 40mph plus. (Alas, the infamous cooking instructions were misquoted; she actually wrote 'first *case* your hare'; 'catch' was far more entertaining and stuck. To 'case' is to strip of skin.) Hare is dark, strong meat; it is also dry meat, hence it is invariably stewed or casseroled rather than roasted. (An exception may be made for the leveret, with its tender, milder flesh.) 'Twas ever thus; Richard II's master cooks prescribed that the 'hayr' be hewed into gobbets, then put in 'cynee', 'papdele' or 'talboytays' – all stews of one kind or another.

The recipe for talboytays included the hare's own 'blode', much like the classic 'jugged' hare, whereby a freshly killed hare is bled into a bowl, its entrails removed, then hung in a larder by its hind legs. Hanging also causes yet more blood to accumulate in the chest cavity. Hares are full of blood; it is haemoglobin that carries the oxygen to their quick limbs. Those who like the 'gameyness' of hare tend to like it very gamey indeed; my farmer grandfather hung his hares till bits of flesh dropped off.

These are the instructions for the making of tal-boytays as given by the master cooks of Richard II in the recipe book *Forme of Cury* (1390).

Take hayrs, and hewe hem to gobbettes, and seeth hem wyth the blode unwaished in brothe of fleshe, and when they buth y-nouh [be enough], cast hem in colde water. Pyke and waish hem clene. Cole the brothe, and drawe it thurgh stynnor [through a strainer].

Take other blode, and cast in boylyng water, seeth it, and draw it thurgh stynnor. Take almanndes un-blanched, waish hem, and grynde hem, and temper it up with the self [same] brothe. Cast al in a pot. Take oynons and parboyle hem. Smyte hem small, and cast hem into the pot, cast thereover powderfort [spiced pepper], vynegar, and salt, temper with wyn, and messe forth.

The kings of England were keen on feasting on hare. For Christmas lunch in 1251 Henry II and his 150 guests consumed 1,300 hares (plus 430 red deer, 200 fallow deer, 200 roe deer, 200 wild swine, 450 rabbits, 2,100 partridges, 290 pheasants, 395 swans, 115 cranes, 400 tame pigs, 70 pork brawns, 7,000 hens, 120 peafowl, 80 salmon, and lampreys 'without number').

The term 'jugging' comes from the practice of cooking an animal inside a jug which is placed in a pan of hot water, a sort of bain-marie for meat. Extravagant metropolitan riffs on the recipe for jugged hare are available. The following is both foolproof and traditional:

Marinate a jointed hare carcass in cider (apple alcohol is a tenderizer), thinly sliced raw onion, black pepper and salt. Leave overnight. Lightly fry the joints to seal them, then place in a casserole, pour in the marinade, plus a teaspoonful of thyme, a squeeze of lemon and a top-up of water. Put in oven on medium heat for 2½ hours. Meanwhile, use the saved blood to make gravy, to which a glass of port or claret is added. Use a slotted spoon to retrieve the hare joints and onion from the casserole. Pour over the gravy.

Expensive to make and prepare, jugged hare was historically a dish fit for a king, or at least minor gentry. In the 1920s the ingredients for jugged hare, excluding the beast itself, came to 6s. 6d., about a day's wages for a farm labourer. A poached hare in the same decade sold for 7s. 6d.

Today, the seeker of jugged hare is likely to have difficulty sourcing a hare, except from specialist game dealers. Scarcity and changes in food taste mean that jugged hare has all but disappeared from the national menu. A survey in 2012 found that hardly any British children knew the dish or, indeed, would wish to eat it.

It is sometimes suggested that the French dish *civet de lièvre* is a Gallic equivalent of the English jugged hare. It is not. *Civet de lièvre* is a stew, cooked on the stove top.

Jugging, stewing or roasting does not exhaust the possibilities of hare meat. Hare pâté and soup were staples in the Victorian and Edwardian country house.

A Victorian Recipe for Hare Soup

A few words ought certainly to be said in acknowledgement of the great merit of hare soups. Of these there are three varieties – the thick, the purée, and the clear. The first is a decoction of hare in good beef stock, thickened without the addition of meat; the second like the first, but with the meat of the hare pounded and blended with it; and the third also made like the first, but, instead of being thickened, clarified, a slight consistency (not to interfere with the clear effect) being imparted by pulverized tapioca or cornflour. Julienne-like strips of the meat of the hare, and similar strips of freshly cooked truffles, form the garnish of this remarkably good soup. Claret should be used in the two first-named soups, chablis in the third. The fame of Scotch hare soup has of course long been established, though there seems to be but little difference between Scotch and English recipes of the old school, and absolutely no difficulty about either. Better flavour may no doubt be obtained from mountain hares ... Large quantities of port wine were formerly considered essential in these preparations, with spices, ham, redcurrant jelly, ketchup, orange or lemon juice, &c. The great Carême, when chef to the Prince Regent, evolved a *potage de lièvre à la St George*, for the moistening of which he required one bottle of the best champagne and one of fine claret, with four ladlefuls of pure consommé. When presented, this intoxicating fluid was garnished with escalopes of hare, quenelles of partridge meat, and mushrooms and truffles in slices. But cultivated taste at the present time looks for simplicity in cookery, and the highly overwrought compounds that were once so popular have passed out of fashion. The chief thing in hare soups now is to have a really good meat stock to cook the hare in – 'brothe of fleshe', as Richard II's cook laid down – to extract the pure flavour of the animal with as few accessories as possible, and to use light French wine in moderate quantity. The back fillets can always be taken for an entree, leaving plenty of material for a good soup.

Colonel Arthur Kenney-Herbert, *The Hare* (1896)

The Hare and the Tortoise

The Hare was once boasting of his speed before the other animals. 'I have never yet been beaten,' said he, 'when I put forth my full speed. I challenge anyone here to race with me.'

The Tortoise said quietly, 'I accept your challenge.'

'That is a good joke,' said the Hare; 'I could dance round you all the way.'

'Keep your boasting till you've beaten,' answered the Tortoise. 'Shall we race?'

So a course was fixed and a start was made. The Hare darted almost out of sight at once, but soon stopped and, to show his contempt for the Tortoise, lay down to have a nap. The Tortoise plodded on and plodded on, and when the Hare awoke from his nap, he saw the Tortoise just near the winning post and could not run up in time to save the race. Then said the Tortoise: '*Plodding wins the race.*'

Aesop (620–564 BC)

IV

The Hare in Myth and Religion

WHEREVER AND WHENEVER the hare has roamed, it has attracted fantasy, above all at springtime and at night. The hare is made for magic. Shy, sighted rarely for much of the year, the hare's astonishing turn of speed only adds to its elusiveness and mystique. As prey animals go, the hare also puts on a convincing aristocratic disdain. In Alison Uttley's Little Grey Rabbit stories the character of Hare is toffee-nosed and picky – an interpretation for children of the animal's natural reserve and an acknowledgement of the hare's status as an exalted beast of venery. It is small surprise that it is Hare who keeps Grey Rabbit up to the mark in housekeeping: '"Where's the milk, Grey Rabbit?" asked Hare. "We can't drink tea without milk."'

In hare mythology and folklore, hares are invested with similar other-worldliness, remoteness, aloofness.

For the Celts, the hare was forbidden flesh, and could not be killed except during the annual ritual hare hunt at Beltane, the Celtic May Day festival. Likewise, the Anglo-Saxons venerated the hare – except during the springtime hare hunt in honour of the goddess Eostre or Ostara, who was depicted as hare-headed. Eostre was the friend of children and her pet white hare laid the brightly coloured eggs of new life, to proclaim the rebirth of the year. Over the centuries Eostre-Hare became transmogrified into the safe and sweet Easter Bunny. But beneath the chocolate-egg veneer of twenty-first-century Easter, the wild echoes of the hare hunt of the Celts and the Anglo-Saxons can still be divined in British folk traditions such as the Hallaton Hare Pie Scramble in Leicestershire on Easter Monday.

The old agricultural name for the April full moon was the Hare Moon. Spring and hares go together like, well, hares and fecundity.

Eostre's name gave us 'Easter', but is also cognate with Old Norse *austr*, meaning 'east', and Greek *eos*, 'dawn', which is another sort of birth, the beginning of a new day. Unrelated ancient cultures the world over associated the hare with dawn and the east. In

Anglo-Saxon 'Englaland' Ostara was the rising sun with the visage of *lepus*, in North American Indian lore Michabo was the Great White Hare, who multi-tasked as the god of the dawn and maker of Earth. Hares are startling at dawn, because the burning sun glows in their eyes.

Some cultures decided that the hare was aphro-disiac, regardless of whether it was spring, summer, autumn or winter. The hare was a companion of the Greek goddess of love, Aphrodite, and the historian

Philostratus adjured that the proper sacrifice to Aphrodite was the hare, as it possessed her virtue of fertility to the nth degree. Heterosexual Greeks carried hare genitals to ward off barrenness; homosexual Greek men presented younger lovers with the gift of a hare.

If the hare is a symbol of spring, of dawn, of fertility, the creature truly belongs to the moon. The ambivalent, inexplicable hare is the lunar animal *sans pareil*. Like the moon, which is always inconstant, the hare is illogical and mysterious. In Ancient Egypt the hare was used as a hieroglyph for the word denoting existence, and all its metaphysical ramifications.

Stare at the full moon and you will see, as tribes the globe over do, shadow-patterns in the shape of a hare. (No, I did not believe it until I tried it one romantic October night, when the wind was wild music in the hedges of a stubble field.) The folklorist Timothy Harley noted in *Moon Lore* (1885):

When the moon is waxing, from about the eighth day to the full, it requires no very vivid imagination to descry on the westward side of the lunar disk a large patch very strikingly resembling a rabbit or hare. The oriental noticing this figure, his poetical

fancy developed the myth-making faculty, which in
process of time elaborated the legend of the hare in
the moon, which has left its marks in every quar-
ter of the globe. In Asia it is indigenous, and is an
article of religious belief. 'To the common people
in India the spots look like a hare, i.e. Chandras,
the god of the moon, carries a hare (sasa), hence
the moon is called Sasin or Sasanka, hare mark or
spot.' Max Müller also writes, 'As a curious coin-
cidence it may be mentioned that in Sanskrit the
moon is called Sasānka, i.e. "having the marks of a
hare," the black marks in the moon being taken for
the likeness of the hare.'

According to Harley, the same hare-in-the-moon
myth travelled to the Hottentots of South Africa,
where it was ornamented to explain the hare's odd lip:

The moon, on one occasion, sent the hare to the
earth to inform men that as she [the moon] died
away and rose again, so mankind should die and rise
again. Instead, however, of delivering this message as
given, the hare, either out of forgetfulness or malice,
told mankind that as the moon rose and died away,
so man should die and rise no more. The hare, having returned to the moon, was questioned as to the

message delivered, and the moon, having heard the true state of the case, became so enraged with him that she took up a hatchet to split his head; falling short, however, of that, the hatchet fell upon the upper lip of the hare, and cut it severely. Hence it is that we see the 'hare-lip'. The hare, being duly incensed at having received such treatment, raised his claws, and scratched the moon's face; and the dark parts which we now see on the surface of the moon are the scars which she received on that occasion.

The hare appears in moon myth north and south, east and west, to such an extent it is a 'lunacy' to match the mad hares of March. For the Norse people, the moon goddess Holda was attended by hares carrying lanterns. The date of Christian Easter is tied to the moon. In Buddhism, when the god Lord Indra was starving, the Buddha in the form of a hare sacrificially leapt into the fire for him, to make a roast ready meal. In return, Lord Indra rewarded the hare by placing his image on the crescent moon. For Buddhists there is a Hare in the Moon rather than a Man in the Moon. Buddhists see the moon-hare as a resurrection symbol, since it is born, grows, reproduces, dies and is reborn.

The Ancient Chinese went one better. In Chinese iconography, the moon-hare holds a pestle and mortar to mix the elixir of immortality. The moon-hare is the giver of eternal life. Down on Earth, female hares, asserted Chinese folklore in startlingly beautiful imagery, are conceived through the touch of the full moon's light (without the need of impregnation by the male), or by traversing moonlit water, or by licking the moonlight from a male hare's coat.

Numerous lunar myths reflect the hare's inexplicableness, its ambiguity. After all, hares do look majestic – but odd, and when they are upright and 'boxing' even oddly human. Kaltes, the moon goddess of western Siberia, shape-shifted into the guise of a hare to roam the hills. In the mythology of the land of the pyramids, hares were intimate with the cycles of the moon, which was masculine when waxing, feminine when waning. By extension, the earthly hare itself moved back and forth between the genders, making it the original gender-fluid being. In Celtic myth Eostre changed shape to become a hare at the full moon; all hares were sacred to her, and acted as her messengers. Small wonder, perhaps, that the Celts were so opposed to eating hare.

Since hares, however, were heavenly messengers, it made sense – to the Celts – that they could be used as instruments of divination, by studying the patterns of their tracks, the rituals of their mating dances, and signs within their entrails. According to the Roman historian

Dio Cassius, Boudicca, queen of the Iceni, finished a war warm-up speech in AD 61, then let 'a hare escape from the fold of her dress; and since it ran on what they considered the auspicious side, the whole multitude shouted with pleasure'. Boudicca further called on the goddess of victory, Andraste, for her support. (Andraste also doubled as the moon goddess.) The poet Robert Graves in *The White Goddess* supposed that the taboo against eating hare in Ancient Britain was due to the fear that killing a hare – a timid beast – might infect the hunter with cowardice. Consequently, he deduced that Boudicca released the hare in the expectation that the Romans might strike at it, and lose their courage.

Under the light of the moon, the hare got up to all sorts of mischief. Hares ambling, then sitting up to detect danger or box or simply perform their toilette looked, thought the people of the past, like a coven of dancing witches. (They do.) During the religiously fanatical years of the Stuarts and the Protectorate in the seventeenth century, witches were widely held to be capable of transforming themselves into hares. At Taunton Assizes in 1663, an old woman called Julian Cox was tried for the practice of sorcery. The court recorded:

A huntsman swore that he went out with a pack of hounds to hunt a hare, and not far from Julian Cox's

house he at last started a hare: the dogs hunted her
very close ... till at last the huntsman perceiving the
hare almost spent and making towards a great bush,
he ran on the other side of the bush to take her up
and preserve her from the dogs; but as soon as he laid
hands on her it proved to be Julian Cox, who had
her head grovelling on the ground and her globes (as
he expressed it) upward. He knowing her, was so af-
frighted that his hair on his head stood on end.

The seventeenth-century witch trials disclosed
the following incantation:

> *I sall gae intill aine haire*
> *With sorrow, and sych, and meikle caire*
> *And I sall gae in the Divellis nam*
> *Ay whill I cam hom againe.*

One Isobel Gowdie, on trial for witchcraft, confessed (under torture) to membership of a coven,
where she transferred her soul into a hare. In 1662
she was burnt at Auldearne. The Scottish 'malkin' or
'mawkin' means both cat and hare, and a common
general name in England for the hare was 'puss'. Hares
are interchangeable with cats in the world of those
who travelled on a broomstick at night. A hare was a

THE PRIVATE LIFE OF THE HARE

common witch 'familiar', or aide. (Hares, it might be noted, are interchangeable with cats in another way: cat milk is a substitute, one of the few, for hare milk when abandoned leverets are raised by human hands.)

A more rational, or at least a more generous, mind than a Cromwellian or Stuart witch-finder general would see that the witch's 'transformation' into a hare was the lingering imprint of the Palaeolithic ceremonies whereby man/woman imitated the animal to be revered, or killed in the hunt, which came to much the same thing.

To revert to human form, the hare-witch required an incantation such as:

> Hare, hare God send thee care.
> I am in a hare's likeness just now,
> But I shall be in a woman's likeness even now.

Witches disguised as hares sucked the milk from cows. The only way of stopping their actions, according to the folklore of Eire, was to kill the witch-hare by shooting it through the heart with a bullet fashioned from a silver coin or a crucifix. In County Kerry in Ireland, well into the twentieth century it was said that eating a hare was like consuming one's own grandmother – perhaps due to the sacred connection between hares and Eostre, or the belief that old 'wise women' could shape-shift into hares by moonlight. In Nordic countries, with the exception of Denmark, witches were believed to create a supernatural animal which resembled a hare and was said to steal milk. The usual name for this animal was *mjolkhare*, 'milkhare'. The milkhare was made from bits of wood, or scrubbing brooms, and was brought to life by the witches' own blood or the touch of a communion wafer.

African hare stories travelled to North America on the slavers' ships, mixed with rabbit tales of the Cherokee and other tribes, and were transformed into the famous Br'er (brother) Rabbit stories of the American South. These stories were passed orally among slaves, for whom Br'er Rabbit was a perfect hero, besting more powerful opponents through his greater intelligence and quicker wits. The Br'er Rabbit stories were written down and published by Joel Chandler Harris in the nineteenth century in a famous collection narrated by the fictional Uncle Remus.

The wiliness of Br'er Rabbit is showcased to perfection in the story whereby Br'er Fox finally traps the leporine anti-hero. Br'er Rabbit tells the fox that he doesn't care what he does to him just as long as he doesn't fling him in the briar patch:

'"Hang me des ez high ez you please, Brer Fox,"
sez Brer Rabbit, sezee, "but do fer de Lord's sake don't
fling me in dat brier-patch," sezee.'

Br'er Rabbit begs the fox to drown him, skin him
– anything but fling him in the briar patch. Of course,
the fox, wanting to give his enemy the worst possible
death, throws him into said briar patch:

Dar wuz a considerbul flutter whar Brer Rabbit
struck de bushes, en Brer Fox sorter hung roun'
fer ter see what wuz gwineter happen. Bimeby he
hear somebody call 'im, en way up de hill he see
Brer Rabbit settin' cross-legged on a chinkapin log
koamin' de pitch outen his har wid a chip. Den
Brer Fox know dat he bin swop off mighty bad. Brer
Rabbit wuz bleedzed fer ter fling back some er his
sass, en he holler out:

'Bred en bawn in a brier-patch, Brer Fox; bred en
bawn in a brier-patch!' en wid dat he skip out des ez
lively ez a cricket in de embers.

In the Native American tradition trickster tales
tend to star Coyote, Jay or Raven but a definite strand,
particularly among the Algonquin-speaking peoples of
the central and eastern woodlands, feature the Hare.
Michabo, the Great Hare, is the bringer of dawn but

also a clown, a lecher, a thief – an amoral figure living on the narrow line between right and wrong.

But the hare always dances dangerously on the edge of things.

The Hare and Christianity

As Christianity took hold in western Europe, hares, so definitely conjoined with the goddess Eostre, came to be seen in unfavourable light – viewed as the aides of witches, or as witches themselves in animal form.

Contrarily, the self-same Christian Church saw virtue in the hare too. The hare, after all, was a fellow victim of persecution, of suffering. In Eadmer's *Life of St Anselm, Archbishop of Canterbury*, there is a telling moment on the road to Hayes in 1097:

Anselm left the court and, while he was hastening to his manor at Hayes the boys of his household with their dogs chased a hare which they came upon in the road. As they were pursuing it, it fled between the feet of the horse on which Anselm sat. The horse stood still; and Anselm – knowing that the wretched animal looked to find a place of refuge beneath him, and not wishing to deny it the help it needed – drew his horse by the reins and kept it still. The dogs came round, snuffling about on all sides and restrained against their will, but they could neither make it move from under the horse, nor harm it in any way.

Anselm forbade the dogs to chase any more. The hare, unhurt, ran off to its fields and wood. Eadmer wrote that the archbishop's retinue was 'not a little uplifted by so affecting a deliverance for the frightened animal'.

There are similarities between Anselm's kindness to the hare and the legend of St Monacella or Melangell, who became the patron saint of hares. Melangell was the daughter of a sixth-century AD Irish king who was determined to marry her to a nobleman. Vowing celibacy, the princess fled across the Irish Sea to Pennant, an isolated spot in the Tanat Valley in North Wales. Here she lived fifteen years without seeing the face of a man. Then one fine day in AD 604 Brochwel Yscythrog, prince of Powys, out hare hunting, pursued his game till he came to a bramble thicket; on entering he was amazed to find a 'maiden of great beauty' engaged in deep devotion. The hare was lying on her robe, boldly facing the dogs. Brochwel urged the dogs, '*Prendite, caniculi, prendite!*' ('Seize her, little dogs, seize her!') but the dogs retreated, howling in fear at the hare. At last the prince addressed Melangell, and heard her story. Greatly impressed, he gave her a parcel of lands, to be a sanctuary to all those that fled there. For all the thirty-seven years of Melangell's retreat 'the wild hares were about her like tame animals'. She was buried in the neighbouring church, called Pennant, and later in her honour Pennant Melangell. The legend of Melangell and the hare is perpetuated by a carving on a fifteenth-century rood screen of hares running to her for protection. Another continuation of the story came in language; hares locally were known as *wyn bach Melangell*, 'St Monacella's Lambs'.

How accurate and true are the stories of Anselm and Melangell is open to question. What is telling, however, is the deliberate use of the hare by the Church to transfer old pagan religion into a Christian context. Albrecht Dürer's woodcut of the Holy Family, *c.* 1496, clearly depicts three hares at the family's feet – the ultimate acceptance and accolade for an animal symbol. (Dürer was, incidentally, a dab hand at portraying hares; his *Feldhase* of 1502 remains the work of hare art by which all others are judged.)

In the same century as Dürer's Holy Family woodcut the hare was consecrated in the very fabric of medieval Christian churches in Europe in the form of triskele hares, a symbol in which three hares are depicted running in a circle and joined by their ears. The triskele is a *trompe l'oeil*. Joined at their tips, the hares' ears form a triangle, each hare apparently with two ears, though the artist has drawn only three in total. The triskele hares in Paderborn Cathedral, Germany, come with an explanatory verse: 'The hares and ears are three. And yet each hare has two (you see).'

Triskele hares are particularly prominent in south-west England, with over thirty recorded examples on roof bosses on the ceilings of churches in Devon, where they are known as 'Tinners' Rabbits'. 'Tinners', of course, were the Devon miners of cassiterite, and 'Rabbits' were hares. Tin mining gave Devon its black mineral wealth; the incorporation of the tri-hare miners' motif in Devon's sacred architecture is a stone proof of the importance of tin to the peninsula. The elders of Devon's churches rendered unto Caesar his due.

What does the three hares iconography stand for in church architecture? The most persuasive explanation is that the intertwined imagery of the three hares was intended to suggest the indivisibility of the Trinity. Additionally, the hare was widely understood to be hermaphrodite, meaning it could reproduce without loss of virginity – like the Virgin Mary. In many locations the three hares symbol is adjacent to a lusty Green Man, providing the congregation with an instructive contrast between the divine and the sinfulness of man.

Triskele hares are not unique to Christianity. The earliest known examples of the design can be found in Buddhist cave temples in China (AD 581–618); from there it spread all along the Silk Road, through the Middle East, Hungary and Poland to Germany, Switzerland and the British Isles. But the Christian Church was always smart on appropriation. After all, the ultimate pagan festival, that of Eostre, became the supreme touchstone of Christianity, the Resurrection at Easter.

Hare Superstitions

The hare can be both lucky or unlucky in superstitions. Hare superstitions, of course, are preserved but corrupted pieces of the ancient beliefs about leporine gods, goddesses and spirits.

- Sailors considered hares so unlucky they could not be mentioned at sea. In Scotland, if the noun 'hare' (or 'pig' or 'rabbit') was spoken aboard a boat, it would be turned around for the shore.

- In Cornish fishing villages, a white hare signalled the coming of a storm.

- A hare crossing the path was unlucky, especially for a pregnant woman, who would miscarry or give birth to a child with a hare-lip. A hare's foot, preferably from the left rear leg, was carried as a charm to avert

71

this; losing the charm would prove very unfortunate. However, a spoken 'apotropaic' charm, such as 'Hare before, Trouble behind: Change ye, Cross, and free me,' to ward off the ill luck could allow the traveller to go on their way in safety. A simpler charm consisted of touching each shoulder with the forefinger, and saying, 'Hare, hare, God send thee care.'

- Alternatively, in the Rhondda Valley of Wales, there is a proverb: 'A witch in the morning, success in the afternoon'; the said 'witch' being a hare. Welsh names for the hare include *cath-hirdaith*, the cat who makes a long journey, *esgair cath*, cat-legged, and *swarngog*, the eared animal.

- A hare's foot, particularly if carried in the right-hand pocket, was said to avert rheumatism and cramps and help actors perform, or was carried as a good luck charm. Samuel Pepys, the diarist, purchased a hare's foot as medicine against a stomach disorder:

  Jan 20th, 1665: So homeward, in my way buying a hare and taking it home, which upon my discourse today with Mr Batten in Westminster Hall, who showed me my mistake that my hare's foote hath not the joynt to it; and assures me he never had his cholique since he carried it about him: and it is a strange thing how fancy works, for I no sooner al-most handled his foote but my belly began to loose and break wind, and whereas I was in some pain yes-terday and tother day, and in fear of more today, I became well, and so continue.

- A hare circling anti-clockwise is unlucky, because it is against the motion of the sun.

- In 1883, in the *Folk-Lore Journal*, William Black wrote, 'From India we learn that it is as unlucky to meet a hare as it is to meet a one-eyed man, an empty water-pot, a carrier without a load, a fox, a jackal, a crow, a widow, or a funeral.'

- A hare's brain taken in wine before bed prevents oversleeping.

- In Cambridgeshire a hare running through the streets is a sign that a fire is about to break out.

- A Cornish superstition maintains that a maiden who dies after being abandoned by her lover will turn into a white hare in order to pursue her faithless love.

- If a farmer sees three or four hares playing together in a 'drove' (the collective name for a group of hares) in March he will have bad luck that year.

- In agricultural lore, the hare has entered into the language of reaping corn. They often hide in corn-fields during the reaping and the last sheaf is often called 'the hare' and its cutting called 'killing the hare', 'cutting the hare' or 'cutting the hare's tail off'. In some places the reapers would all stand around and throw their sickles at the 'hare'.

- Naming the hare was thought to bring bad luck. One anonymous thirteenth-century poem lists seventy-seven alternative names for 'Sir Hare', the creature that must not be named.

The hare, the hare-kin,
Old Big-bum, Old Bouchart,
The hare-ling, the frisky one,
Old Turpin, the fast traveller,
The way-beater, the white-spotted one,
The lurker in ditches, the filthy beast,
Old Wimount, the coward,
The slink-away, the nibbler,
The one it's bad luck to meet, the white-livered,
The scutter, the fellow in the dew,
The grass nibbler, Old Goibert,
The one who doesn't go straight home, the traitor,
The friendless one, the cat of the wood,

The starer with wide eyes, the cat that lurks in the broom,
The purblind one, the furze-cat,
The clumsy one, the blear-eyed one,
The wall-eyed one, the looker to the side,
And also the hedge-frisker,
The stag of the stubble, long-eared,
The animal of the stubble, the springer,
The wild animal, the jumper,
The short animal, the lurker,
The swift-as-wind, the skulker,
The shagger, the squatter in the hedge,
The dew-beater, the dew-hopper,
The sitter on its form, the hopper in the grass,
The fidgety-footed one, the sitter on the ground,
The light-foot, the sitter in the bracken,
The stag of the cabbages, the cropper of herbage,
The low creeper, the sitter-still,
The small-tailed one, the one who turns to the hills,
The get-up quickly,
The one who makes you shudder,
The white-bellied one,
The one takes refuge with the lambs,
The numbskull, the food mumbler,
The niggard, the flincher,
The one who makes people flee, the covenant-breaker,
The snuffler, the cropped head

(His chief name is Scoundrel),
The stag with the leathery horns,
The animal that dwells in the corn,
The animal that all men scorns,
The animal that no one dare name.

The Witch of Easington

British and Irish folk tales are replete with accounts of men led astray by hares who are really witches in disguise, or of hares revealed as women (usually, though not always, crones) when they are wounded in their animal shape. In one well-known story from Dartmoor, a mighty hunter named Bowerman disturbed a coven of witches practising their rites, and so one young witch determined to take revenge upon the man. She shape-shifted into a hare, led Bowerman through a deadly mire, then turned the hunter and his hounds into piles of stones, which can still be seen today. (The stone formations are known by the names Hound Tor and Bowerman's Nose.) The Celtic warrior Oisin, one of the Fianna band of heroes, was equally luckless in his encounter with a hare-woman. Out hunting on horseback, his hound chased a hare and took a bite out of its flank. The hare sought refuge in a clump of

rushes. When Oisin followed the hare he found a door
leading into the ground. He eventually emerged into
a huge hall, where he found a beautiful young woman
sitting on a silver throne, bleeding from a wound in her
leg. Oisin fell in love with her, but when he wanted
to go back into the world, she told him he had been
with her in her palace for seven hundred years. He did
not believe her, but as soon as he reached the earth he
instantly became a very old man.

In 'The Witch of Easington', published in *Legends
and Superstitions of the County of Durham* (1886),
William Brockie recorded a variant of one of the most
enduring, widespread legends from hare mythology,

whereby the hunter wounds a hare who is transmuted into her real being, a witch.

Mrs. Mary Shaw, who died about three years ago, at the age of eighty-five, and who went to live at Easington when she was forty years old, was told by the elderly people of that old village that, in their young days, whenever the neighbouring gentry went out with the harriers to hunt over any of the farms round about Easington or Castle Eden, it always occurred that a hare started up and carried the dogs off the right scent, straight towards the former place, and somehow or other, without any of the usual doublings and windings, always managed to throw them out and get clear away. This happened so often that it was plain to be seen that there was something uncanny about this crafty member of the leporine genus. Somebody at length suggested that it must certainly be a witch, for witches, according to common credence, often take out-door exercise, in the form of such fleet creatures; and it was noticed that a certain cottage in the place used to be shut up on the days when the sports were held, as if its solitary inmate, an ill-natured wrinkled old hag, had gone abroad somewhere. She usually worked, indeed, in the fields, so that this need not have been wondered at, but her sour temper and ill habits had rendered her hateful to all her neighbours, so it was noways unnatural that suspicion should fall upon her to the purport that she was a witch, and consequently the identical mysterious hare.

In order to test the truth of this, the master of the hounds was advised to get a black bloodhound, which must have been suckled at a woman's breast, and set it on the uncanny creature's track next time it appeared, then he was assured that its capture would be certain. A hound answering this description was accordingly got, and next field day it led the hunt. The hare had never been so closely followed up before. It made, as usual, direct for Easington, but instead of the hound being thrown off the scent, it kept up the pursuit until the old woman's cottage door was reached. A little hole had been cut in that door, for the hens to go in and out at. The hare rushed forward to get through the hole, but the black hound was too close behind to let it get in unscathed. Just as it was darting through, he caught it on the haunch, and tore away a bit of the flesh. The huntsmen hurried up, and, finding the door fast barred, they burst it open.

On entering they saw one of the strangest sights that ever human eye was set on. No hare was to be seen, nor any other living brute beast, but there sat

79

the witch, bathed in sweat and shivering in agony, with the blood streaming from her on to the floor. The poor creature, we are told, confessed her guilt, which indeed, she could not easily have denied under the circumstances; and she earnestly begged pardon and asked to be forgiven. Her charm had been broken by the drawing of the blood, and her power was henceforth gone, even if she had wished to exercise it. The gentlemen charitably took pity on her, and left her there alone, to staunch her wound as she best knew how. Never after that were the Easington and Castle Eden harriers thrown off the legitimate scent by any such diabolic means.

The Garden

Well – one at least is safe.
One shelter'd hare
Has never heard the sanguinary yell
Of cruel man, exulting in her woes.
Innocent partner of my peaceful home,
Whom ten long years' experience of my care
Has made at last familiar; she has lost
Much of her vigilant instinctive dread,
Not needful here, beneath a roof like mine. Yes – thou
May'st eat thy bread, and lick the hand
That feeds thee; thou may'st frolic on the floor
At evening, and at night retire secure
To thy straw couch, and slumber unalarm'd;
For I have gain'd thy confidence, have pledg'd
All that is human in me to protect
Thine unsuspecting gratitude and love.
If I survive thee I will dig thy grave;
And, when I place thee in it, sighing, say,
I knew at least one hare that had a friend.

William Cowper (1731–1800)

V

The Pet Hares of William Cowper

WILLIAM COWPER, THE eighteenth-century writer and poet, kept three hares as pets in his home in Olney, Buckinghamshire. When Cowper moved to Olney, in 1768 at the age of thirty-seven, he had already suffered several mental breakdowns, and the hares entered his life to ease dark days. The hares must have given relief if not cure, since he wrote what is widely considered his best work, 'The Task', while at Olney.

Cowper wrote a long letter about his pet hares to *The Gentleman's Magazine* in 1794. In it Cowper explained that in 1774 he took in a leveret that had been given to a neighbour's children and was not thriving; he thought that caring for the animal might help his mental state, and so it did. He was soon gifted two other leverets:

I undertook the care of three, which it is necessary that I should here distinguish by the names I gave them, Puss, Tiney, and Bess. Notwithstanding the two feminine appellatives, I must inform you that they were all males. Immediately commencing carpenter, I built them houses to sleep in; each had a separate apartment so contrived that their ordure would pass thro' the bottom of it; an earthen pan placed under each received whatsoever fell, which being duly emptied and washed, they were thus kept perfectly sweet and clean. In the daytime they had the range of a hall, and at night retired each to his own bed, never intruding into that of another.

Tiney was never tamed, Puss 'grew presently familiar', and would allow himself to be what one might call a 'lap hare', petted and carried around. Bess proved the tamest of the three:

Bess had a courage and confidence that made him tame from the beginning. I always admitted them into the parlour after supper, when the carpet affording their feet a firm hold, they would frisk and bound and play a thousand gambols, in which Bess, being remarkably strong and fearless, was always superior to the rest, and proved himself the Vestris of the party. [August Vestris, 1760–1842, was a celebrated French dancer.] One evening the cat being in the room had the hardiness to pat Bess upon the cheek, an indignity which he resented by drumming upon her back with such violence, that the cat was happy to escape from under his paws and hide herself.

To Cowper's surprise, the hares were wholly discerning in their attitude to visitors. Some they were never reconciled to, while the miller 'engaged their affections at once', since his flour-covered coat 'had charms that were irresistible'. Cowper's close acquaintance with Tiney, Puss and Bess taught him to 'hold the sportsman's amusement in abhorrence', for he 'little knows what amiable creatures he persecutes', or 'how cheerful they are in their spirits, what enjoyment they have of life'.

Cowper kept careful notes on the hares' preferred foodstuffs. 'Sowthistle, dandelion, and lettuce were their favourite vegetables, especially the last.' He discovered by accident when cleaning out a birdcage that they liked white sand ('I suppose as a digestive'), and they accounted green corn, or maize, 'a delicacy, both blade and stalk, but the ear they seldom eat'. They also liked wheat straw, aromatic herbs, caned musk. To prevent scouring or diarrhoea he fed the hares squares of bread, 'placed it every evening in their chambers, for they feed only at evening and in the night'.

Bess died soon after he was full grown. Tiney lived to be nine years old, and was the subject of a verse eulogy, 'Epitaph on a Hare':

> Here lies, whom hound did ne'er pursue,
> Nor swifter greyhound follow,
> Whose foot ne'er tainted morning dew,
> Nor ear heard huntsman's hallo',
>
> Old Tiney, surliest of his kind,
> Who, nurs'd with tender care,
> And to domestic bounds confin'd,
> Was still a wild Jack-hare.
>
> Though duly from my hand he took
> His pittance ev'ry night,
> He did it with a jealous look,
> And, when he could, would bite.
>
> His diet was of wheaten bread,
> And milk, and oats, and straw,
> Thistles, or lettuces instead,
> With sand to scour his maw.
>
> On twigs of hawthorn he regal'd,
> On pippins' russet peel;

And, when his juicy salads fail'd,
Slic'd carrot pleas'd him well.

A Turkey carpet was his lawn,
Whereon he lov'd to bound,
To skip and gambol like a fawn,
And swing his rump around.

His frisking was at evening hours,
For then he lost his fear;
But most before approaching show'rs,
Or when a storm drew near.

Eight years and five round-rolling moons
He thus saw steal away,
Dozing out all his idle noons,
And ev'ry night at play.

I kept him for his humour's sake,
For he would oft beguile
My heart of thoughts that made it ache,
And force me to a smile.

But now, beneath this walnut-shade
He finds his long, last home,

And waits in snug concealment laid,
'Till gentler Puss shall come.

He, still more agèd, feels the shocks
From which no care can save,
And, partner once of Tiney's box,
Must soon partake his grave.

Puss was almost twelve when he died on 9 March 1786, 'of mere old age, and apparently without pain'.

Cowper and his hares became so well known that Olney has a weathervane in the market place bearing a quill pen. And a hare.

Hare Proverbs

- It's hard to put a grain of salt on a hare's tail

- When the hare dies the fox mourns (there but for the grace of God ...)

- To kiss the hare's foot (to be late; the hare has gone by and left its footprint for you to salute)

- Who hunteth two hares loses the one and leaveth the other

- Hare horns, tortoise hairs (i.e. rare)

- Even a hare will bite when it is cornered (China)

- Even a hare will insult a dead lion (Latin)

- Deer-hunter, waste not your arrow on the hare

- 'Blood more stirs/To rouse a lion than to start a hare' (Shakespeare)

- Nature will come through the claws, and the hound will follow the hare

- Find yourself without an excuse, and find a hare without a smeuse (a hare's gateway)

- A hare or a clump of ferns (for two things that are indistinguishable)

- You cannot run with the hare and hunt with the hounds

- Even the rustling of leaves will alarm the hare

- Little dogs start the hare, the great get her

VI

Hare Today

THE HARE IS one of the immemorial, defining British animals. When Vita Sackville-West (1892–1962) wanted to invoke the timeless pastoralism of Britain in her bestselling novel of 1930, *The Edwardians*, what natural items did she list? Well, 'the verdure of the trees, the hares and the deer'.

William Wordsworth (1770–1850) made much the same literary point about the centrality of the hare to our countryside in 'Resolution and Independence' (1802):

All things that love the sun are out of doors;
The sky rejoices in the morning's birth;
The grass is bright with rain-drops; – on the moors
The hare is running races in her mirth;
And with her feet she from the plashy earth
Raises a mist, that glittering in the sun,
Runs with her all the way, wherever she doth run.

The war poet Wilfrid Wilson Gibson (1878–1962) wrote a long poem, 'The Hare', which caught in its lines the uplift many get from the sight of hare in a field:

> Upon my belly in the straw,
> I lay, and watched her sleek her fur,
> As, daintily, with well-licked paw,
> She washed her face and neck and ears:
> Then, clean and comely in the sun,
> She kicked her heels up, full of fun,
> As if she did not care a pin
> Though she should jump out of her skin,
> And leapt and lolloped, free of fears,
> Until my heart frisked round with her.

When Gibson wrote 'The Hare' there were about four million brown hares in Britain. The Victorian-Edwardian countryside was likely the best of all natural worlds for the brown hare. The 'patchwork quilt' landscape consisted of a mix of cereals, root crops and grass for livestock. Fields were relatively small. Under the traditional ley rotation the sequence of cereals was followed by grass. There was, accordingly, forage and cover within an easy lope the year long. As late as 1946 Brian Vesey-Fitzgerald could write in *British Game* that there 'are far too many hares'. Hares, he added, were 'a menace to agriculture', especially in East Anglia, which overran with hares. Ten hares, Norfolk farmers estimated, ate as much as a sheep; west Suffolk farmers considered the beet crop was reduced by as much as two tons per acre by the browsing of *lepus*.

Recent surveys show the brown hare has declined by more than 60 per cent during the past sixty years. In some parts of Britain, such as the South-West, the brown hare is now a rarity, even locally extinct. (The Ground Game Act of 1880 gave tenant farmers the right to kill hares and rabbits on their farms in order to protect their crops; particular eradications of the brown hare in the West Country may date from this time.)

Unlike rabbits, hares are resistant to myxomatosis and have suffered no equivalent genocidal viral cull. So, what accounts for the hare's decline? As with so many forms of British wildlife, hares have been devastated by changing agricultural practice. Hares do not hibernate or store useful amounts of fat, so they need a continuous and nutritious food supply. Since the Second World War haymaking has largely been replaced by silage production; grass for silage tends to be sown to a single species, resulting in a 'bio-uniformity' which almost certainly explains the scarcity of hares in western areas where dairy and beef farming predominates. (To make matters worse for hares, they dislike pastures with the high densities of livestock typical of intensive farming.) Hares now fare better in the arable areas of East Anglia, giving a distinct east–west divide in the leporine national population. Indeed, in the three bastion counties of Norfolk, Cambridgeshire and Suffolk, brown hare numbers seem stable.

Other changes in the pattern of land use have hindered hares. The big switch from spring-sown to autumn-sown cereal crops means hares have an abundant food supply between November and February … after which the plants become unpalatable. Effectively, there is famine for hares when energy needs are greatest – peak breeding time. In the absence of varied grass and palatable cereals, hares have to eke out grazing along hedges. Alas, hares have been deprived of 150,000 miles of hedgerow since the 1960s.

Hares, especially in their infantile form, are vulnerable to being killed by farm machinery, since they will sit immobile as the blades of grass-mowing machinery go over them, or the boom from a crop-sprayer passes overhead. Back in the bad old days of the 1960s, when pesticide/herbicide/molluscicide/fungicide use was cruder, the corpses of ten, twenty, thirty hares could be found in an arable field in the time after spraying. DDT and paraquat were then farmers' favourites, and hares ingested the poisons when they licked themselves clean.

Something else has changed in the countryside. Many predators are now more abundant than they were. That's you, Mr Fox. Foxes rarely, if ever, surprise and kill adult hares, yet they can systematically prey on leverets to cause extinction of the local hare population. A single fox family can consume eighty leverets a year.

Whammy upon whammy. Climate change, with increased summer rainfall, will harm the brown hare. Under wet conditions, breeding success is poor, with leverets succumbing to diseases such as coccidiosis.

Despite its decline, the hare is the only game species in Britain which does not have a shooting close season. The Hares Preservation Act 1892 merely prohibits the sale of hares or leverets between 1 March and 31 July. (Hare must not be on the menu in restaurants during this period.) Large, organized shoots in East Anglia during February and March can account for 40 per cent of the entire national brown hare population. And since the breeding season is well under way by February, orphaned leverets are left to die of starvation. The preternaturally fecund hare can recover from such slaughter but only on a level, biodiverse and sympathetic running field.

Like the brown hare, the Irish hare has gone into long-term decline due to changed farming practices.

The Irish hare's killing for sport may also have had a major impact. The Ulster Wildlife Trust has warned that hare hunting and coursing 'may prove to be the final straw for some of the more isolated populations'. Organized hare coursing has now been banned in Northern Ireland but remains widespread in the Irish Republic. Throughout Ireland, the Irish hare remains a quarry species and may be hunted with guns or dogs despite the evidence of decline. The data suggests that hare numbers have dropped in Northern Ireland by around 50 per cent over the last decade.

As for the mountain hare of Scotland, the picture also has bleak edges. The current number of mountain hares in Scotland is unclear but the latest research suggests a decline of 43 per cent since 1995, and virtual extinction in some parts where it was previously abundant. Excessive grazing by deer, sheep and cattle has depleted the heather so that less food and cover is available for the hares. That said, population densities of mountain hares in the Highlands tend to fluctuate every seven to ten years, over the course of which hare density can change tenfold to reach numbers in excess of 250 hares per square kilometre. The reason for the fluctuation is related to parasite burdens. The mountain hare is affected by tick-borne louping-ill virus and a gut parasite, *Trichostrongylus retortaeformis*, which dramatically reduces fecundity. The tick-borne virus kills grouse chicks, so mountain hares are snared and shot on grouse moors to maintain the grouse-shooting industry. The Scottish Parliament has instigated a close season for the mountain hare (1 March to 31 July).

We still have hares today, but they may be gone tomorrow, and we will be the losers, because hares are the life and soul of the British landscape. To see a hare sit still as stone, to watch a hare boxing on a frosty March morning, to witness a hare bolt ... these are great things. Every field should have a hare.

Hare Conservation Dos and Don'ts for Farmers and Country Dwellers

- Hares need quiet, undisturbed cover for raising leverets. On livestock farms, leave some areas of grass uncut and ungrazed till late in the season. Or better still, leave field corners and buffer strips ungrazed, unfertilized and only cut every five years.

- When making silage, cut from the inside of the field outwards, so that hares can escape from the machinery into adjacent fields.

- Hares like a 'patchwork quilt' farmland. On arable farms, break up large blocks of cereal as much as possible.

- Plant game cover and wildflower strips – or lawns – to create cover and food for hares. Although hares prefer open land they will graze country gardens.

- Hares should not be shot in late winter unless crops are being severely damaged.

- Poaching and hare coursing are illegal (the latter was banned under the 2004 Hunting Act) and should be reported to the police Wildlife Liaison Officer.

Useful Organizations and Addresses

Game & Wildlife Conservation Trust www.gwct.org.uk

Hare Preservation Trust www.hare-preservation-trust.co.uk

Joint Nature Conservation Committee http://jncc.defra.gov.uk

The Wildlife Trusts www.wildlifetrusts.org

Envoy: A Hare Limerick

There was an old man whose despair
Induced him to purchase a hare:
Whereon one fine day,
He rode wholly away,
Which partly assuaged his despair.

Edward Lear (1812–88)

ACKNOWLEDGEMENTS

As ever, there are thanks due, because no book is an island. So, thank you, Susanna Wadeson, Lizzy Goudsmit, Josh Benn, Ella Horne, Sophie Christopher, Julian Alexander, David Blanchard and Penny Lewis-Stempel.

PICTURE CREDITS